THE POETRY OF PROMETHIUM

The Poetry of Promethium

Walter the Educator

Silent King Books a WhichHead Imprint

Copyright © 2023 by Walter the Educator

All rights reserved. No part of this book may be reproduced in any manner whatsoever without written permission except in the case of brief quotations embodied in critical articles and reviews.

First Printing, 2023

Disclaimer
This book is a literary work; poems are not about specific persons, locations, situations, and/or circumstances unless mentioned in a historical context. This book is for entertainment and informational purposes only. The author and publisher offer this information without warranties expressed or implied. No matter the grounds, neither the author nor the publisher will be accountable for any losses, injuries, or other damages caused by the reader's use of this book. The use of this book acknowledges an understanding and acceptance of this disclaimer.

"Earning a degree in chemistry changed my life!"
- Walter the Educator

dedicated to all the chemistry lovers, like myself, across the world

CONTENTS

Dedication	v
Why I Created This Book?	1
One - Symbol Of Knowledge	2
Two - Promethium, A Legend	4
Three - Celestial Delight	6
Four - Jewel Of The Divine	8
Five - Eternal Youth	10
Six - Forever Renowned	12
Seven - Mystery To Explore	14
Eight - Annals Of Science	16
Nine - Fearless And Bold	18
Ten - Captivates The Heart	20
Eleven - Yast And Immense	22
Twelve - Boundless Human Quest	24

Thirteen - Driven By The Unknown	. . . . .	26
Fourteen - Depths Of Your Allure	. . . . .	28
Fifteen - Endless Romance	. . . .	30
Sixteen - Never Tire	. . . .	32
Seventeen - Gracefully Twirl	. . . . . .	34
Eighteen - Sparks Our Fascination	. . . . .	36
Nineteen - Whisper In The Dark	. . . . .	38
Twenty - Promethium, A Puzzle	. . . . .	40
Twenty-One - Forever Stark	. . . . . .	42
Twenty-Two - Secrets Reside	. . . . . .	44
Twenty-Three - Can't Tame	. . . . . .	46
Twenty-Four - Atomic Core	. . . . . .	48
Twenty-Five - We Persist	. . . .	50
Twenty-Six - Catalyst For Discovery	. . . .	52
Twenty-Seven - Rearrange	. . . .	54
Twenty-Eight - Hold In Our Hands	. . . .	56
Twenty-Nine - Uncover The Truths	. . . .	58
Thirty - Piece By Piece	. . . .	60
Thirty-One - Fog Of Uncertainty	. . . . .	62
Thirty-Two - Inspiring Us All	. . . . . .	64

Thirty-Three - Humanity's Maze 66

Thirty-Four - Promethium, A Gem 68

About The Author 70

WHY I CREATED THIS BOOK?

Creating a poetry book about the chemical element Promethium was an intriguing and unique endeavor. Promethium, with its fascinating properties and historical significance, offers a wealth of inspiration for poetic exploration. This book delves into various themes associated with Promethium, such as its discovery, atomic structure, uses, and symbolic representations. Through poetry, I can capture the essence of Promethium and explore its scientific, philosophical, and metaphorical dimensions. This endeavor allows for a fusion of science and art, providing a platform to appreciate the beauty and complexity of the element while engaging readers on multiple levels.

ONE

SYMBOL OF KNOWLEDGE

In the realm of elements, a hidden gem I find,
Promethium, a mystery, a treasure undefined.
Within the periodic table's vast expanse,
It shines with a radiance that makes hearts dance.
    Promethium, oh enigmatic element,
With atomic number sixty-one, heavens sent.
Its existence in nature is hard to find,
Yet scientists pursue it with an ardent mind.
    A phantom of the elements, elusive and rare,
Its creation in labs requires utmost care.
Through the fusion of atoms, a flicker ignites,
Revealing Promethium's mystical delights.
    Its glow, a beacon in the darkest night,
A luminescent spectacle, a wondrous sight.

Captivating the curious, it beckons with grace,
Promethium, a puzzle for the human race.
   In the realm of science, it sparks curiosity,
A puzzle piece in the grand cosmic tapestry.
Its properties, its uses, still largely unknown,
Promethium, a riddle waiting to be shown.
   So let us marvel at this elemental quest,
And delve into Promethium's secrets, at best.
For in the world of elements, it stands apart,
A symbol of knowledge, a beacon of art.

TWO

PROMETHIUM, A LEGEND

In the realm of elements, a hidden tale unfolds,
Promethium, a legend waiting to be told.
A phantom presence, mysterious and rare,
With atomic number sixty-one, it dares.

A ghostly essence, shimmering in the night,
Promethium, a captivating light.
Born from atomic fusion's fiery dance,
It beckons seekers with an alluring glance.

Forged in the heart of stars, a celestial birth,
Promethium, a treasure of infinite worth.
Its secrets locked within its atomic core,
A riddle that scientists strive to explore.

With properties enigmatic, yet to be unveiled,
Promethium, a wonder that has prevailed.

Its luminescent glow, a spectral embrace,
Guiding wanderers through time and space.
   A symbol of potential, of endless intrigue,
Promethium, a catalyst for minds to besiege.
It fuels our yearning for knowledge and lore,
A testament to humanity's thirst to explore.
   So let us marvel at this enigmatic prize,
Promethium, a mystery that defies.
In the vast universe, it stands alone,
A testament to the wonders yet unknown.
   In the realm of elements, it claims its throne,
Promethium, a legend to be forever known.

THREE

CELESTIAL DELIGHT

In the cosmos' symphony, a note yet unsung,
Promethium, a tale that must be spun.
A whisper in the void, a clandestine dance,
This enigmatic element, a cosmic romance.

With sixty-one protons, it claims its place,
Promethium, a luminary in space.
Born in the heart of stars, its birth divine,
A radiant jewel, in the celestial design.

Through fusion's fervor, it's brought to life,
Promethium, a luminescent strife.
But elusive it remains, a phantom in the night,
Defying capture, evading human sight.

Its secrets lie hidden, a cryptic enigma,
Promethium, a puzzle that ignites the stigma.

In laboratories, minds strive to reveal,
The nature of this element, its true appeal.
    A beacon of curiosity, it draws the brave,
Promethium, an element that seeks to pave,
New paths of knowledge, uncharted and vast,
A testament to the power of questions asked.
    Oh Promethium, with your spectral hue,
You captivate minds, as the universe grew.
A symbol of wonder, of infinite might,
Promethium, a star that guides us through the night.
    In the cosmic tapestry, your story unfolds,
Promethium, a legend that forever holds,
A place in our quest for truth and light,
Promethium, a celestial delight.

FOUR

JEWEL OF THE DIVINE

In the realm of elements, a secret to explore,
Promethium, a mystery we can't ignore.
With sixty-one protons, it claims its name,
A radiant jewel in the scientific game.

Promethium, oh elusive and rare,
A tale untold, a cosmic affair.
Born in dying stars, a celestial spark,
Revealing its essence in the laboratory dark.

Beyond the periodic table's known terrain,
Promethium's allure, an enigmatic refrain.
A dance of electrons, a celestial waltz,
Unveiling its secrets, the mind's somersaults.

In labs, scientists toil with fervent might,
Cracking the code of Promethium's light.

Its glowing embers, a captivating sight,
Illuminating the path to scientific height.
   A phantom element, veiled in disguise,
Promethium, a treasure beneath distant skies.
With uses untapped, potential unknown,
Promethium, a realm we have yet to own.
   Oh Promethium, a cosmic enigma,
A puzzle unsolved, a riddle enigma.
We delve into your depths, seeking to find,
The wonders and secrets you've left behind.
   In the grand tapestry of elements, you shine,
Promethium, a jewel of the divine.
A testament to human curiosity and might,
Promethium, a beacon in the quest for light.

FIVE

ETERNAL YOUTH

Promethium, the phantom of the realm,
An element mysterious, at science's helm.
Born from the stars, a celestial birth,
It dances in shadows, a treasure unearthed.

In the depths of atoms, its secrets reside,
A puzzle unsolved, where answers hide.
Scientists strive, with minds ablaze,
To unravel the mysteries, in countless ways.

Promethium's glow, a spectral embrace,
Illuminates the path, through time and space.
A radiant beacon, in the darkest night,
Guiding seekers, with its ethereal light.

Yet elusive it remains, a fleeting dream,
Promethium, an enigma, or so it seems.
Its properties unique, its uses unknown,
A symphony of elements, yet to be shown.

Oh Promethium, a cosmic enigma,
A mystery that captivates, like an enchanter's stigma.
We yearn to understand your atomic core,
To unlock the secrets, to know you more.

In the grand tapestry of the periodic table,
Promethium, a fragment, an enigmatic fable.
A testament to the limits of our knowledge,
A reminder to keep seeking, to keep the edge.

So we embrace the challenge, the quest for truth,
Promethium, a symbol of eternal youth.
In the realm of elements, you stand apart,
A cosmic riddle, etched on every scientist's heart.

SIX

FOREVER RENOWNED

Promethium, a whisper in the dark,
A clandestine element, leaving its mark.
In the hidden corners of the periodic table,
It dances with mystery, an enigmatic fable.
   Its atomic number, fifty-nine, a clue to its might,
Promethium, a cosmic puzzle, shining so bright.
Yet its existence fleeting, as it swiftly decays,
Leaving scientists in awe, in an eternal maze.
   Oh Promethium, an elusive apparition,
A tantalizing secret, a scientific mission.
With its radioactive glow, a captivating sight,
It beckons us to explore, to search for its light.
   Through the depths of uncertainty, we strive,
To unravel the secrets Promethium does hide.

In laboratories, minds race to comprehend,
The essence of this element, a quest without end.

Promethium, a celestial enigma so grand,
A testament to human curiosity, hand in hand.
As we unravel the universe's cosmic code,
Promethium stands proud, a symbol of the unknown.

In the vast expanse of scientific endeavor,
Promethium, a treasure we tirelessly endeavor.
Though its presence is scarce, its impact profound,
Promethium, a mystery forever renowned.

SEVEN

MYSTERY TO EXPLORE

Promethium, an elusive flame ablaze,
In the realm of elements, a rare maze.
With atomic secrets that it guards,
A mystery that science regards.

Within the periodic table's domain,
Promethium's allure remains untamed.
Its radioactive glow, a mesmerizing sight,
Captivating minds with its ethereal light.

A symbol of curiosity, it beckons and calls,
Scientists' endeavors, it enthralls.
In laboratories, experiments unfold,
Seeking the truths Promethium holds.

Oh Promethium, a puzzle unsolved,
Yet its potential, we shall evolve.
Its applications, a realm unexplored,
Waiting for minds to be restored.

In the vast cosmos of elements unknown,
Promethium, a star that's yet to be shown.
A testament to human pursuit,
Promethium, a symbol of resolute.

So let us delve into its atomic core,
Promethium, a mystery to explore.
In the tapestry of science, it shines,
An enigma that forever defines.

EIGHT

ANNALS OF SCIENCE

Promethium, a captivating allure,
A chemical enigma, forever obscure.
Its presence elusive, its secrets concealed,
Promethium, a mystery yet to be revealed.
 Within the realm of the periodic table it resides,
A rare element that continuously hides.
Scientists endeavor to understand its essence,
To grasp the nature of its luminescence.
 Oh Promethium, a spectral enigma so rare,
With your atomic puzzle, we're left in a stare.
A symbol of the unknown, of uncharted realms,
Promethium, a beacon that overwhelms.
 In laboratories, minds tirelessly toil,
Seeking the answers through endless trial.
Promethium, a puzzle we yearn to solve,
To unravel the enigma that it involves.

In the cosmic tapestry of elements untamed,
Promethium, a celestial gem unclaimed.
A testament to the human thirst for knowledge,
To explore the universe, to venture beyond the edge.

So we persist in our quest, with passion ablaze,
Promethium, a mystery that forever stays.
In the annals of science, your story unfolds,
Promethium, a legend waiting to be told.

NINE

FEARLESS AND BOLD

Promethium, a clandestine essence, rare and true,
A hidden gem in the cosmic brew.
With atomic allure, it whispers in the night,
Promethium, a mystery, shrouded in its own light.
    In laboratories, minds embark on a quest,
To unravel the secrets, to put it to the test.
Promethium, a puzzle waiting to be solved,
A riddle for the curious, a mystery to be evolved.
    Oh Promethium, with your enigmatic charm,
Scientists' hearts race, like a beating alarm.
Exploring your properties, your atomic grace,
Promethium, a captivating element to embrace.
    In the symphony of elements, you stand alone,
Promethium, a masterpiece, yet to be fully known.

A testament to the vastness of the universe's store,
Promethium, a symbol of exploration and more.
  So we journey into the depths, fearless and bold,
Promethium, a treasure waiting to unfold.
In the realm of science, your legacy will reside,
Promethium, a quest we'll forever abide.

TEN

CAPTIVATES THE HEART

Promethium, an element of intrigue,
In the realm of science, a mystique.
Its name derived from Prometheus' fire,
Igniting curiosity, a burning desire.

A rare entity with atomic might,
Promethium, shining in scientific light.
Radiating energy, a captivating glow,
Unleashing secrets, the cosmos bestow.

Oh Promethium, a cosmic enigma,
Unraveling your mysteries, a scientific stigma.
In the depths of laboratories, minds explore,
The wonders you hold, the secrets in store.

A phantom element, elusive and scarce,
Promethium, a puzzle beyond compare.

With each discovery, new questions arise,
A perpetual quest that never dies.
  In the tapestry of elements, you stand tall,
Promethium, a symbol of knowledge's call.
A testament to human ingenuity and might,
Promethium, a beacon in the quest for light.
  So we delve into your atomic core,
Promethium, a mystery we yearn to explore.
In the grand symphony of the universe's art,
Promethium, a melody that captivates the heart.

ELEVEN

VAST AND IMMENSE

Promethium, a name that echoes through time,
An element rare, with a story sublime.
In the realm of the periodic table, you reside,
Promethium, an enigma we cannot hide.

With an atomic dance, you emit a glow,
A captivating light that sparks the show.
Radiating energy, in the realm of the unseen,
Promethium, a wonder that fascinates the keen.

Scientists delve into your atomic code,
Unraveling the secrets that you hold.
Promethium, an intrigue in the laboratory's domain,
A puzzle to solve, a prize to gain.

As we unlock your mysteries, we find,
Promethium, a substance that's one of a kind.
A symbol of exploration, of human quest,
Promethium, a treasure to manifest.

In the grand tapestry of elements unknown,
Promethium, a gem that truly has shone.
A reminder that the universe is vast and immense,
Promethium, a marvel that leaves us in suspense.

TWELVE

BOUNDLESS HUMAN QUEST

Promethium, a radiant enigma, elusive and rare,
In the realm of elements, a celestial affair.
With atomic whispers and a glowing hue,
You captivate scientists with wonders anew.

A dance of electrons, a symphony of light,
Promethium, a spectacle that ignites.
In laboratories, minds endeavor to unveil,
The secrets you hold, the stories you entail.

Oh Promethium, a phantom of the periodic chart,
A symbol of exploration and scientific art.
In the vast cosmic expanse, you reside,
A testament to the mysteries we strive to confide.

As we peer into your atomic core,
Promethium, a treasure we tirelessly explore.

Unlocking the secrets, one by one,
Revealing the essence of this element, undone.
　In the grand tapestry of elements untold,
Promethium, a legend waiting to unfold.
A reminder of the boundless human quest,
To unravel the universe's eternal bequest.

THIRTEEN

DRIVEN BY THE UNKNOWN

In the realm of elements, a hidden gem,
Promethium, a mystery, a rare emblem.
Glowing with a radiant hue, ever so bright,
A captivating element, a celestial light.

In laboratories, scientists immerse,
Seeking the truths that you disperse.
Promethium, a puzzle, a quest to unfold,
Unraveling its properties, yet untold.

Oh Promethium, a symbol of intrigue,
The enigma in your atomic league.
A phantom element, elusive and rare,
Challenging minds, a scientific affair.

In the tapestry of the periodic table,
Promethium, a chapter that we enable.

A testament to human curiosity and might,
Promethium, a beacon in the quest for light.
  So we delve deeper, driven by the unknown,
Promethium, a journey we're eager to own.
In the vast cosmic dance, you play your part,
Promethium, forever etched in science's heart.

FOURTEEN

DEPTHS OF YOUR ALLURE

Promethium, a spark of luminescent glow,
Hidden in the depths where few dare to go.
A mystical element, elusive and rare,
Promethium, a secret waiting to be shared.
    In laboratories, scientists seek your trace,
Unveiling your properties, with patient grace.
Promethium, a puzzle they strive to solve,
Unlocking the mysteries that you involve.
    Oh Promethium, a symbol of innovation,
A catalyst for scientific exploration.
With each discovery, the world is amazed,
Promethium, a marvel that won't be erased.
    In the grand tapestry of the atomic domain,
Promethium, a gem that continues to reign.

A testament to humanity's thirst for knowledge,
Promethium, a flame that forever will burn.
    So we journey into the depths of your allure,
Promethium, a treasure we constantly adore.
In the vast universe, your presence is known,
Promethium, a story waiting to be shown.

FIFTEEN

ENDLESS ROMANCE

Promethium, a beacon in the dark abyss,
A mystery waiting to be unraveled, such bliss.
In laboratories, scientists toil and strive,
To understand your essence, to bring you alive.

Oh Promethium, an element of intrigue,
With properties that make minds intrigue.
Radiating energy, a luminescent glow,
Promethium, a spectacle to behold.

In the realm of atoms, you hold your place,
Promethium, a symbol of cosmic grace.
A puzzle unsolved, a riddle unexplained,
Promethium, a challenge that keeps minds engaged.

As we delve into your atomic core,
Promethium, a treasure we truly adore.
Unlocking your secrets, one by one,
Revealing the wonders that you've spun.

In the grand tapestry of the periodic chart,
Promethium, a jewel that sets us apart.
A testament to human curiosity and might,
Promethium, a catalyst for scientific light.
So we journey into the depths of your allure,
Promethium, a quest that will endure.
In the vast expanse of the universe's expanse,
Promethium, a symbol of our endless romance.

SIXTEEN

NEVER TIRE

Promethium, a hidden gem of lore,
Radiating with an enigmatic allure.
In the realm of elements, you hold your ground,
A mystical substance yet to be fully found.

Oh Promethium, a phantom in the atomic sea,
With properties that defy what we foresee.
Scientists strive to unravel your mystery,
To grasp the essence of your chemistry.

In laboratories, minds dance with delight,
Exploring the depths of your atomic might.
Promethium, a symbol of constant quest,
A journey into the unknown, we're blessed.

As we peer into your atomic core,
Promethium, a treasure we forever adore.
Unveiling the secrets that rest within,
Revealing the wonders we're destined to win.

In the grand tapestry of the scientific realm,
Promethium, a chapter that overwhelms.
A reminder of the vastness of our exploration,
Promethium, a beacon of endless fascination.

So we venture forth, driven by desire,
Promethium, a flame that will never tire.
To discover the truths that you hold,
Promethium, a story yet to be told.

SEVENTEEN

GRACEFULLY TWIRL

Promethium, a dance of atomic fire,
A captivating element, our hearts inspire.
In the realm of science, you claim your place,
Promethium, a symbol of cosmic grace.

Mysterious and rare, you beckon us near,
Promethium, a puzzle we're eager to clear.
With each electron spin, a tale unfolds,
Promethium, a wonder that science beholds.

In the tapestry of elements, you're a gem,
Promethium, a catalyst for knowledge to stem.
Unraveling your secrets brings a thrill,
Promethium, a journey we're destined to fulfill.

Oh Promethium, a radiant light in the night,
A symbol of curiosity, shining so bright.

In laboratories, minds yearn to explore,
Promethium, a key to unlock the unknown door.
    So we venture forth, with passion ablaze,
Promethium, a source of wonder and praise.
In the cosmic dance, you gracefully twirl,
Promethium, a marvel that makes our spirits unfurl.

EIGHTEEN

SPARKS OUR FASCINATION

Promethium, a mystic element unseen,
In the vast universe, a cosmic dream.
A phantom of the periodic chart,
Promethium, a puzzle to tear apart.

Oh Promethium, a dance of atomic might,
A symbol of radiance in the darkest night.
With each electron spinning in its shell,
Promethium, a tale that science will tell.

In the grand tapestry of the atomic realm,
Promethium, a secret at the helm.
A testament to our relentless quest,
Promethium, a challenge we won't rest.

So we delve into your atomic lore,
Promethium, our curiosity soars.

Unlocking the secrets that you hold,
Promethium, a story yet to be told.
　In laboratories, minds are ablaze,
Promethium, a beacon in our scientific maze.
As we seek to understand your atomic core,
Promethium, a mystery we adore.
　Promethium, your presence sparks our fascination,
A symbol of discovery and exploration.
In the cosmic symphony, you play your part,
Promethium, forever etched in science's heart.

NINETEEN

WHISPER IN THE DARK

Promethium, a whisper in the dark,
A luminescent spark, a celestial arc.
In the realm of elements, you stand alone,
Promethium, a mystery yet to be known.

In laboratories, minds tirelessly seek,
Promethium, a puzzle for the curious and meek.
Unraveling your secrets, a scientific dance,
Promethium, a symbol of our relentless chance.

Oh Promethium, elusive and rare,
A captivating tale, we yearn to share.
In the vast expanse of the atomic sea,
Promethium, a pearl awaiting its decree.

So we journey forth, guided by light,
Promethium, a beacon shining bright.

Exploring the depths of your atomic core,
Promethium, a riddle we forever adore.
 Promethium, a chapter waiting to be penned,
A testament to human minds that transcend.
In the tapestry of science, you leave your mark,
Promethium, a celestial spark in the dark.

TWENTY

PROMETHIUM, A PUZZLE

Promethium, a secret in the night,
A shimmering star, a captivating sight.
In the realm of elements, you stand alone,
Promethium, a mystery yet to be known.
 Like a phoenix rising from the ash,
Promethium, a symbol of resilience and clash.
In the atomic dance, you hold your ground,
Promethium, a treasure waiting to be found.
 Through the currents of time, you've journeyed far,
Promethium, a celestial wanderer, a shining star.
Unveiling your essence, a scientific quest,
Promethium, a puzzle we're determined to digest.
 Oh Promethium, a catalyst for change,
In laboratories, minds rearrange.

Unlocking your secrets, we strive to see,
Promethium, a bridge to the unknown, set free.
    So we venture forth, with hearts afire,
Promethium, our driving desire.
In the vast cosmic expanse, you reside,
Promethium, a symbol of knowledge, our guide.

TWENTY-ONE

FOREVER STARK

Promethium, a hidden spark in the night,
A tale of intrigue, a beacon of light.
In the realm of elements, you stand alone,
Promethium, a mystery yet to be fully known.
    Through the waves of time, you quietly reside,
Promethium, a secret we cannot hide.
Unveiling your secrets, a scientific chase,
Promethium, a puzzle we eagerly embrace.
    Oh Promethium, an enigma of the atomic plane,
A symbol of wonder, a mystery to explain.
In the depths of laboratories, minds ignite,
Promethium, a source of intellectual delight.
    So we embark on a journey, with curiosity ablaze,
Promethium, a path only science can pave.

Seeking answers to your atomic dance,
Promethium, a captivating, elusive trance.
   Promethium, a testament to human endeavor,
A symbol of our thirst for knowledge, forever.
In the tapestry of elements, you leave your mark,
Promethium, a spark of curiosity, forever stark.

TWENTY-TWO

SECRETS RESIDE

Promethium, a whisper in the night,
A chemical element, shining so bright.
Within your atomic heart, secrets reside,
Promethium, a mystery we cannot hide.

In laboratories, minds wander and dream,
Promethium, a puzzle to unravel, it seems.
With every experiment, a step closer we tread,
Promethium, a path to knowledge we spread.

Oh Promethium, a captivating allure,
A symbol of science's relentless explore.
Through the vast expanse of the periodic table,
Promethium, a tale we're eager to enable.

In the cosmic symphony, you play your part,
Promethium, a melody that touches the heart.

A catalyst for change, a spark of innovation,
Promethium, a source of scientific inspiration.
    So we embrace the challenge you present,
Promethium, a quest that won't relent.
In the realm of elements, you shine so bright,
Promethium, a beacon guiding us through the night.

TWENTY-THREE

CAN'T TAME

Promethium, an element rare and unseen,
In the realm of atoms, you reign supreme.
A mystery shrouded in atomic lore,
Promethium, a puzzle we strive to explore.
    Oh Promethium, your glow seduces the eye,
A beacon of radiation that lights up the sky.
In laboratories, scientists pursue,
Promethium, a quest to unravel the truth.
    Through the depths of the periodic table we roam,
Promethium, a treasure we yearn to call home.
Unraveling your secrets, a scientific delight,
Promethium, a challenge that ignites our insight.
    Promethium, a symbol of progress and change,
In the pursuit of knowledge, we rearrange.

Your atomic dance, a captivating sight,
Promethium, a symbol of scientific might.
 So we delve deeper, driven by curiosity's flame,
Promethium, a quest we can't tame.
In the cosmic tapestry, you leave your mark,
Promethium, a celestial spark in the dark.

TWENTY-FOUR

ATOMIC CORE

Promethium, an element rare and true,
A mystery hidden, waiting to debut.
In the realm of atoms, you hold your place,
Promethium, an enigma we embrace.

Oh Promethium, with atomic allure,
A testament to scientific endeavor.
Through the trials of discovery, we seek,
Promethium, an answer we eagerly peek.

In laboratories, minds ignite,
Promethium, a puzzle we strive to light.
Unveiling your secrets, a tantalizing task,
Promethium, a challenge we dare not mask.

Promethium, a beacon in the night,
A symbol of progress, shining so bright.
In the vast expanse of the chemical sphere,
Promethium, a whisper we long to hear.

So we delve into your atomic realm,
Promethium, a treasure at the helm.
Unlocking the doors to your atomic core,
Promethium, a secret forevermore.

TWENTY-FIVE

WE PERSIST

Promethium, a mystical name,
In the realm of elements, you claim your fame.
With atomic secrets, you captivate,
Promethium, a puzzle we anticipate.
    In the depths of laboratories, minds ignite,
Promethium, a subject of scientific delight.
Unraveling your mysteries, we aspire,
Promethium, a quest that fuels our fire.
    Oh Promethium, with atomic glow,
A symbol of knowledge, a treasure to bestow.
Through the periodic table, we traverse,
Promethium, a discovery we immerse.
    In the cosmic symphony, you play your role,
Promethium, a melody that touches the soul.

A catalyst for progress, a beacon of light,
Promethium, a symbol of scientific might.
    So we venture forth, with minds open wide,
Promethium, a journey we can't hide.
In the quest for understanding, we persist,
Promethium, a puzzle we can't resist.

TWENTY-SIX

CATALYST FOR DISCOVERY

Promethium, a mystic chemical tale,
In the realm of elements, a legend set to unveil.
With atomic allure, you captivate our minds,
Promethium, a puzzle that constantly binds.

In laboratories, scientists persist,
Promethium, a challenge they cannot resist.
Through trials and tests, they seek to find,
Promethium, a treasure to unlock and bind.

Oh Promethium, a symbol of curiosity's flame,
In the pursuit of knowledge, we stake our claim.
Your atomic dance, a mesmerizing sight,
Promethium, a beacon of scientific light.

Through the vast expanse of time and space,
Promethium, a secret we strive to embrace.

In the tapestry of elements, you shine so bright,
Promethium, a symbol of scientific might.
 So we venture forth, with fervor and zeal,
Promethium, a quest that's both grand and surreal.
In the realm of chemistry, you hold the key,
Promethium, a catalyst for discovery.

TWENTY-SEVEN

REARRANGE

Promethium, a hidden gem of the periodic table,
A story of science that's yet to be fable.
In the depths of knowledge, you reside,
Promethium, a mystery we can't hide.

Oh Promethium, with atomic allure,
A symbol of progress that we must secure.
Through the boundless cosmos, we embark,
Promethium, a journey that leaves a mark.

In laboratories, minds alight,
Promethium, a puzzle we strive to excite.
Unveiling your secrets, a quest we pursue,
Promethium, a challenge we never eschew.

Promethium, a spark of innovation and change,
In the realm of elements, you rearrange.

Your atomic dance, an enigmatic sway,
Promethium, a symbol of scientific play.
 So we dive into the depths of the unknown,
Promethium, a treasure we're yet to own.
With curiosity as our guiding light,
Promethium, a beacon shining so bright.

TWENTY-EIGHT

HOLD IN OUR HANDS

Promethium, a mystery in the realm of the small,
Unveiling your secrets, we heed the call.
In laboratories, minds eagerly delve,
Promethium, a story we strive to unravel.

    A symbol of ingenuity, you shine,
Promethium, a jewel in the scientific line.
Through the periodic table, we wander,
Promethium, a quest we ponder.

    Promethium, a catalyst for change and reaction,
Igniting progress, a scientific satisfaction.
In the depths of atoms, your essence resides,
Promethium, a beacon that guides.

    So we explore the depths of your atomic core,
Promethium, a journey we can't ignore.

With every discovery, our knowledge expands,
Promethium, a testament to human hands.

Promethium, a testament to human hands.
In the pursuit of truth, we bravely stand.
In laboratories, we strive to understand,
Promethium, a puzzle we hold in our hands.

Promethium, a puzzle we hold in our hands.
Seeking answers, we traverse unknown lands.
With every experiment, our understanding expands,
Promethium, a story that forever stands.

TWENTY-NINE

UNCOVER THE TRUTHS

Promethium, an element of intrigue,
With atomic secrets, you intrigue.
In the realm of science, you hold sway,
Promethium, a puzzle we aim to portray.

Through the waves of uncertainty, we sail,
Promethium, a journey that will never fail.
Unraveling your mysteries, one by one,
Promethium, a quest that's just begun.

Promethium, a spark within the dark,
A symbol of knowledge, a striking mark.
In laboratories, minds ignite,
Promethium, a beacon of scientific light.

So we delve into your atomic domain,
Promethium, a treasure we hope to attain.
With each discovery, our understanding grows,
Promethium, a catalyst that constantly shows.

Promethium, a shimmering cosmic jewel,
Unveiling your secrets, a task we never rule.
In the vast expanse of the chemical sea,
Promethium, a symbol of curiosity's plea.

So we pursue you, Promethium, with zest,
To uncover the truths you possess.
In the realm of elements, you stand tall,
Promethium, a wonder that enthralls us all.

THIRTY

PIECE BY PIECE

Promethium, a name whispered with awe,
A chemical enigma, waiting to thaw.
In the depths of the periodic table you reside,
Promethium, a secret we can't hide.

Radiant and rare, your atomic dance,
Captivating scientists with each glance.
A symbol of progress, of relentless pursuit,
Promethium, a puzzle that keeps us astute.

Through the fog of uncertainty, we press on,
Promethium, a challenge we won't abandon.
In laboratories, minds ignite,
Promethium, a beacon of scientific light.

Unveiling your mysteries, piece by piece,
Promethium, a quest that'll never cease.
In the realm of elements, you hold your place,
Promethium, a testament to human grace.

So we unravel the secrets you keep,
Promethium, a journey that's not for the meek.
With every experiment, we come closer to know,
Promethium, a story that continues to grow.

THIRTY-ONE

FOG OF UNCERTAINTY

In the realm of elements, a hidden gem,
Promethium, a mysterious emblem.
With atomic allure, you beckon and call,
Promethium, a puzzle standing tall.
    In the depths of science, we seek your trace,
Promethium, a challenge we embrace.
Through the fog of uncertainty, we press on,
Promethium, a quest that's never withdrawn.
    Promethium, a spark of curiosity's flame,
Whispering secrets, a tantalizing game.
In laboratories, minds ignite,
Promethium, a beacon shining bright.
    Unraveling your mysteries, we strive,
Promethium, a journey that makes us thrive.

With each discovery, our knowledge expands,
Promethium, a puzzle in our hands.
  Promethium, a symbol of scientific might,
Guiding us through the depths of the night.
In the cosmic symphony, you play your part,
Promethium, a treasure close to the heart.

THIRTY-TWO

INSPIRING US ALL

Promethium, a whisper in the dark,
A mystery that ignites the spark.
In the realm of elements, you reside,
Promethium, a secret we can't hide.

Through the depths of time and space,
Promethium, a puzzle we chase.
In laboratories, minds alight,
Promethium, a beacon shining bright.

Your atomic dance, a rhythmic glide,
Promethium, a symphony inside.
Unveiling your secrets, a quest we pursue,
Promethium, a journey that feels so new.

Promethium, a symbol of endless quest,
A treasure we must uncover and wrest.
In the tapestry of science, you hold a place,
Promethium, a reminder of our human grace.

So we venture forth, with passion and might,
Promethium, a quest that feels so right.
In the pursuit of knowledge, we stand tall,
Promethium, inspiring us all.

THIRTY-THREE

HUMANITY'S MAZE

In the depths of the periodic table, you reside,
Promethium, a mystery we can't hide.
An element rare, elusive and bright,
Promethium, a captivating light.

Promethium, a tale of atomic allure,
A dance of electrons, so uncertain and pure.
In laboratories, minds ignite,
Promethium, a quest for scientific insight.

Promethium, a symbol of endless pursuit,
A puzzle we strive to unravel, to loot.
With every experiment, we seek to know,
Promethium, a story that continues to grow.

Promethium, a spark in the darkest of nights,
Guiding us through the unknown, with celestial might.
In the realm of elements, you hold a unique space,
Promethium, a secret we long to embrace.

So we venture on, with curiosity ablaze,
Promethium, a chapter in humanity's maze.
In the pursuit of knowledge, we steadfastly roam,
Promethium, a beacon that leads us home.

THIRTY-FOUR

PROMETHIUM, A GEM

Promethium, a whisper in the dark,
A mystery that ignites a spark.
In laboratories, minds enthralled,
Promethium, a tale waiting to be recalled.
  In the depths of atoms, you reside,
Promethium, a secret we confide.
Unveiling your essence, we strive,
Promethium, a puzzle that keeps us alive.
  Promethium, an enigma we pursue,
A quest that unveils something new.
Through the waves of uncertainty, we sail,
Promethium, a beacon that will never fail.
  Promethium, a symbol of exploration,
A catalyst for scientific inspiration.

In the realm of elements, you stand alone,
Promethium, a gem that has shone.
 So we delve deeper, with hearts afire,
Promethium, a desire we can't tire.
With every discovery, our awe increases,
Promethium, a wonder that never ceases.

ABOUT THE AUTHOR

Walter the Educator is one of the pseudonyms for Walter Anderson. Formally educated in Chemistry, Business, and Education, he is an educator, an author, a diverse entrepreneur, and he is the son of a disabled war veteran. "Walter the Educator" shares his time between educating and creating. He holds interests and owns several creative projects that entertain, enlighten, enhance, and educate, hoping to inspire and motivate you.

Follow, find new works, and stay up to date
with Walter the Educator™
at WaltertheEducator.com

www.ingramcontent.com/pod-product-compliance
Lightning Source LLC
LaVergne TN
LVHW052001060526
838201LV00059B/3767